ADDITIONAL SKILL AND DRILL MANUAL

CATHY FERRER
Valencia Community College

TRIGONOMETRY

NINTH EDITION

Margaret L. Lial
American River College

John Hornsby
University of New Orleans

David I. Schneider
University of Maryland

AF807830

PEARSON
Addison Wesley

Boston San Francisco New York
London Toronto Sydney Tokyo Singapore Madrid
Mexico City Munich Paris Cape Town Hong Kong Montreal

Reproduced by Pearson Addison-Wesley from electronic files supplied by the author.

Copyright © 2009 Pearson Education, Inc.
Publishing as Pearson Addison-Wesley, 75 Arlington Street, Boston, MA 02116.

All rights reserved. No part of this publication may be reproduced, stored in a retrieval system, or transmitted, in any form or by any means, electronic, mechanical, photocopying, recording, or otherwise, without the prior written permission of the publisher. Printed in the United States of America.

ISBN-13: 978-0-321-53052-3
ISBN-10: 0-321-53052-7

2 3 4 5 6 OPM 10 09 08

PEARSON
Addison
Wesley

Trigonometry

Contents

Find the complement and the supplement of each angle.

1. 80° 2. 10° 3. 40° 4. 9°

Find the measure of each angle.

5. supplementary angles with measures $10x$ and $5x$ degrees

6. supplementary angles with measures $6x + 4$and $2x$ degrees

7. complementary angles with measures $7x - 9$ and $7x + 15$ degrees

8. complementary angles with measures $10x$ and $20x$ degrees

Perform each calculation.

9. $135°16'45'' + 234°50''$

10. $41°47'42'' - 32°15'9''$

11. $26°47' - 31°10'$

12. $42°46'15'' + 25°7'37''$

Convert each angle measure to decimal degrees. Round to the nearest thousandth of a degree.

13. $26°30'$

14. $66°54'3''$

15. $10°42'5''$

16. $121°51'22''$

Convert each angle measure to degrees, minutes, and seconds.

17. 73.2°

18. 48.27°

19. 83.41°

20. −36.78°

Find the angle of smallest positive measure coterminal with each angle.

21. −60°

22. 470°

23. 1500°

24. −123°

Give an expression that generates all angles coterminal with each angle. Let *n* represent any integer.

25. 60°

26. 120°

The measures of two angles of a triangle are given. Find the measure of the third angle.

1. $34°$, $15°$

2. $47°37'49''$, $26°53'22''$

3. $26.8°$, $93.2°$

Solve each problem.

4. Sherri, who is 63 inches tall, casts a shadow that is 95 inches long. At the same time, Tony's shadow is 101 inches long. How tall is Tony?

5. An apartment building casts a shadow that is 40 feet long at the same time one of the tenants casts a shadow of 8 feet. If the tenant is 5.5 feet tall, how tall is the apartment building?

6. A 6-foot person is standing next to a light pole. The person's shadow is 2.5 feet long and the light pole's shadow is 50 feet. How tall is the light pole?

7. A tree casts a shadow that is 20 meters long at the same time an upright 1-meter stick casts a shadow that is 2.75 meters long. How tall is the tree?

8. Two triangles are similar. The lengths of the three sides of the larger triangle are 5, 7, and 10 inches. The longest side of the other triangle is 4 inches. Find the other two sides of the smaller triangle.

9. A pair of vertical angles have measures of $(4x - 9)°$ and $(2x + 3)°$ respectively. Find the measure of each angle.

10. A pair of vertical angles have measures of $(5x - 7)°$ and $(3x + 9)°$ respectively. Find the measure of each angle.

11. The three angles of a triangle measure $(2x - 4)°$, $(7x - 2)°$, and $(x + 3)°$ respectively. Find the measure of each angle.

12. The three angles of a triangle measure $(x - 3)°$, $(6x - 4)°$, and $(3x + 8)°$ respectively. Find the measure of each angle.

13. The three angles of a triangle measure $3x°$, $3x°$, and $3x°$ respectively. Find the measure of each angle.

14. Two quadrilaterals are similar. The lengths of the two longest sides of the first quadrilateral are 40 inches and 60 inches. The lengths of the three shortest sides of the second quadrilateral are 5 inches, 12 inches and 16 inches. Find the unknown lengths of the sides of these two figures.

Find the values of the six trigonometric functions for angles in standard position having the following point on their terminal side.

1. $(4,-3)$
2. $(-6,0)$
3. $\left(-\sqrt{3},1\right)$
4. $(-1,-1)$
5. $\left(3,-\sqrt{7}\right)$
6. $(9,40)$
7. $(-3,-4)$

In the following exercises, an equation of the terminal side of an angle θ in standard position is given with a restriction on x. Find the values of the six trigonometric functions of θ.

8. $-12x+5y=0;\ x\geq 0$
9. $-x+y=0;\ x\leq 0$
10. $40x+9y=0;\ x\geq 0$
11. $-5x+12y=0;\ x\geq 0$
12. $5x+3y=0;\ x\leq 0$

Use the trigonometric function values of quadrantal angles to evaluate each expression.

13. $\sin^2 90°+\cos 180°\tan 0°$
14. $\cos 90°\sin 90°-\tan^2 0°$
15. $\csc^2 90°\sin^2 270°$
16. $\cos 0°+\sec 0°\sin 90°$
17. $\sin 90°\sin 0°+\sin 180°\sin 360°$
18. $\cos^2 180°\sec^2 180°$
19. $\sin^2 90°+\cos^2 90°$
20. $\sin 90°\cos 0°+\cos 90°\sin 0°$

Use the appropriate reciprocal identity to find each function value.

1. $\sin\theta$ if $\csc\theta = -2$
2. $\cos\theta$ if $\sec\theta = \sqrt{11}$
3. $\tan\theta$ if $\cot\theta = \frac{\sqrt{6}}{2}$

Determine the signs of the trigonometric functions of an angle in standard position with the given measure.

4. 36°
5. 112°
6. 310°

Find the value of each variable.

7. $\sin(5\theta - 4)° = \frac{1}{\csc(8\theta - 4)°}$
8. $\tan(6\theta)° = \frac{1}{\cot(2\theta + 4)°}$

Identify the quadrant for the angle satisfying the given conditions.

9. $\sin\theta < 0,\ \tan\theta < 0$
10. $\csc\theta > 0,\ \cot\theta < 0$
11. $\sec\theta < 0,\ \sin\theta > 0$

Decide whether each statement is possible or impossible for an angle θ.

12. $\csc\theta = 0.5$
13. $\tan\theta = -6$
14. $\cos\theta = 1$

Use identities to find each function value.

15. $\tan\theta$, if $\cos\theta = -\frac{4}{5}$, with in quadrant III
16. $\cos\theta$, if $\cot\theta = -1$, with in quadrant II
17. $\sin\theta$, if $\tan\theta = 2$, with in quadrant I

Find all the trigonometric function values for each angle.

18. $\cos\theta = -\frac{8}{17}$, with in quadrant II
19. $\sec\theta = \frac{13}{5}$, with in quadrant IV
20. $\sin\theta = -\frac{1}{2}$, with in quadrant IV

1.1 Angles

1. 10°, 100°

2. 80°, 170°

3. 50°, 140°

4. 81°, 171°

5. 120°, 60°

6. 136°, 44°

7. 33°, 57°

8. 60°, 30°

9. 369°17′35″

10. 9°32′33″

11. –4°23′

12. 67°53′52″

13. 26.500°

14. 66.901°

15. 10.701°

16. 121.911°

17. 73°12′

18. 48°16′12″

19. 83°24′36″

20. –36°46′48″

21. 300°

22. 110°

23. 60°

24. 237°

25. $60° + 360n°$

26. $120° + 360n°$

1.2 Angle Relationships and Similar Triangles

1. 131°

2. 15°28′49″

3. 60°

4. 66.98 inches

5. 2525 feet

6. 120 feet

7. 7.27 meters

8. 2, 2.8 inches

9. 15°

10. 33°

11. 32.6°, 126.1°, 21.3°

12. 14.9°, 103.4°, 61.7°

13. 60°, 60°, 60°

14. 12.5, 24, 30 inches

1.3 Trigonometric Functions

In exercises 1-12 we give, in order, sine, cosine, tangent, cotangent, secant, and cosecant.

1. $-\frac{3}{5}; \frac{4}{5}; -\frac{3}{4}; -\frac{4}{3}; \frac{5}{4}; -\frac{5}{3}$

2. 0; –1; 0; undefined; –1; undefined

3. $\frac{1}{2}; -\frac{\sqrt{3}}{2}; -\frac{\sqrt{3}}{3}; -\sqrt{3}; -\frac{2\sqrt{3}}{3}; 2$

4. $-\frac{\sqrt{2}}{2}; -\frac{\sqrt{2}}{2}; 1; 1; -\sqrt{2}; -\sqrt{2}$

5. $-\frac{\sqrt{7}}{4}$; $\frac{3}{4}$; $-\frac{\sqrt{7}}{3}$; $-\frac{3\sqrt{7}}{7}$; $\frac{4}{3}$; $-\frac{4\sqrt{7}}{7}$

6. $\frac{40}{41}$; $\frac{9}{41}$; $\frac{40}{9}$; $\frac{9}{40}$; $\frac{41}{9}$; $\frac{41}{40}$

7. $-\frac{4}{5}$; $-\frac{3}{5}$; $\frac{4}{3}$; $\frac{3}{4}$; $-\frac{5}{3}$; $-\frac{5}{4}$

8. $\frac{12}{13}$; $\frac{5}{13}$; $\frac{12}{5}$; $\frac{5}{12}$; $\frac{13}{5}$; $\frac{13}{12}$

9. $-\frac{\sqrt{2}}{2}$; $-\frac{\sqrt{2}}{2}$; 1; 1; $-\sqrt{2}$; $-\sqrt{2}$

10. $-\frac{40}{41}$; $\frac{9}{41}$; $-\frac{40}{9}$; $-\frac{9}{40}$; $\frac{41}{9}$; $-\frac{41}{40}$

11. $\frac{5}{13}$; $\frac{12}{13}$; $\frac{5}{12}$; $\frac{12}{5}$; $\frac{13}{12}$; $\frac{13}{5}$

12. $\frac{5\sqrt{34}}{34}$; $-\frac{3\sqrt{34}}{34}$; $-\frac{5}{3}$; $-\frac{3}{5}$; $-\frac{\sqrt{34}}{3}$; $\frac{\sqrt{34}}{5}$

13. 1 **14.** 0 **15.** 1 **16.** 2

17. 0 **18.** 1 **19.** 1 **20.** 1

1.4 Using the Definitions of the Trigonometric Functions

In exercises 18-20 we give, in order, sine, cosine, tangent, cotangent, secant, and cosecant.

1. $-\frac{1}{2}$

2. $\frac{\sqrt{11}}{11}$

3. $\frac{\sqrt{6}}{3}$

4. All are positive

5. Sine and Cosecant are positive; all others are negative.

6. Tangent and Cotangent are positive; all others are negative.

7. 0

8. 1

9. IV **10.** II **11.** II **12.** impossible

13. possible **14.** possible

15. $\frac{3}{4}$

16. $-\frac{\sqrt{2}}{2}$

17. $\frac{2\sqrt{5}}{5}$

18. $\frac{15}{17}$; $-\frac{8}{17}$; $-\frac{15}{8}$; $-\frac{8}{15}$; $-\frac{17}{8}$; $\frac{17}{15}$

19. $-\frac{12}{13}$; $\frac{5}{13}$; $-\frac{12}{5}$; $-\frac{5}{12}$; $\frac{13}{5}$; $-\frac{13}{12}$

20. $-\frac{1}{2}$; $\frac{\sqrt{3}}{2}$; $-\frac{\sqrt{3}}{3}$; $-\sqrt{3}$; $\frac{2\sqrt{3}}{3}$; -2

Suppose ABC is a right triangle with sides of lengths *a, b,* and *c* with right angle at C. Find the unknown side length using the Pythagorean theorem, and then find the values of the six trigonometric functions for angle B. Rationalize denominators when applicable.

1. $a = 4,\ c = 5$

2. $a = 8,\ b = 15$

Write each function in terms of its cofunction. Assume that all angles in which an unknown appears are acute angles.

3. $\sin 16°$

4. $\tan 48.27°$

5. $\cos(\alpha + 75°)$

6. $\sin(\theta - 19°)$

Find one solution for each equation. Assume that all angles in which an unknown appears are acute angles.

7. $\sin\theta = \cos(\theta + 10°)$

8. $\tan(2\theta - 14°) = \cot(\theta + 20°)$

9. $\cos(90° + \theta) = \sin(2\theta - 9°)$

10. $\tan(3\theta + 6°) = \cot(4\theta)$

Determine whether the following inequalities are true or false.

11. $\sin 29° \geq \sin 39°$

12. $\cos 46° \leq \cos 44°$

13. $\tan 75° \leq \tan 80°$

14. $\sin 23° \geq \cos 23°$

For each expression, give the exact value.

15. $\sin 45°$

16. $\tan 45°$

17. $\sin 60°$

18. $\csc 60°$

19. $\sec 60°$

20. $\cot 60°$

Find exact values of the six trigonometric functions for each angle. Rationalize denominators when applicable.

1. $135°$

2. $390°$

3. $-690°$

4. $-120°$

5. $225°$

6. $-420°$

7. $1380°$

8. $-930°$

Find the exact value of each expression.

9. $\sin 480°$

10. $\cos 510°$

11. $\tan(-1950°)$

Determine whether each statement is true or false.

12. $\sin 60° = 2\sin 30° \cos 30°$

13. $\cos^2 135° - \sin^2 135° = 1$

14. $\cos 90° = \cos 60° + \cos 30°$

Suppose θ is in the interval (90°,180°). Find the sign of each of the following.

15. $\tan\frac{\theta}{2}$

16. $\sin(\theta + 90°)$

17. $\cos(\theta + 90°)$

Find all values of θ, if θ is in the interval [0°, 360°) and has the given function value.

18. $\sin\theta = \frac{1}{2}$

19. $\csc\theta = -\frac{2\sqrt{3}}{3}$

20. $\cot\theta = \sqrt{3}$

Use a calculator to find a decimal approximation for each value. Give as many digits as your calculator displays.

1. $\sin 4.62°$
2. $\cos 10.59°$
3. $\tan 204°$
4. $\cot 2.56°$
5. $\sec 59.724°$
6. $\csc 185°15'$
7. $\sin 16°57'13''$
8. $\cos 25°38'8''$
9. $\tan 132°14'9''$
10. $\csc 78°41'24''$

Find a value of θ in the interval [0°, 90°] that satisfies each statement. Leave answers in decimal degrees.

11. $\sin\theta = .46947156$
12. $\cos\theta = .89284798$
13. $\tan\theta = .73457303$
14. $\csc\theta = 1.15470054$

Use a calculator to evaluate each expression.

15. $\cos^2 49° + \sin^2 49°$
16. $\cos 95.64° \cos 84.36° - \sin 95.64° \sin 84.36°$

Use a calculator to decide whether each statement is true or false.

17. $\sin 80° = 2\sin 40° \cos 40°$
18. $\cos 25° + \cos 25° = \cos 50°$
19. $\sin 84° = 2\cos^2 42° - 1$
20. $\cos^2 62° = 1 - \sin^2 62°$

Solve each right triangle. In each case, $C = 90°$.

1. $A = 30°$, $b = 100$

2. $B = 35°$, $b = 5$

3. $A = 32.6°$, $a = 41.6$

4. $A = 50°$, $c = 12$

5. $B = 40°$, $c = 9.1$

6. $B = 32.54°$, $a = 6.7$

7. $B = 52°$, $a = 31$

8. $A = 75°$, $c = 20$

9. $A = 68°$, $b = 80$

10. $b = 21.12$, $c = 52.25$

11. $B = 30°$, $a = 40$

12. $a = 15$, $b = 20$

13. $B = 38°$, $a = 15$

14. $c = 6.587$, $a = 5.476$

15. $b = 120$, $c = 140$

16. $a = 28.4$, $c = 51.3$

17. $A = 60°$, $c = 22$

18. $B = 25.4°$, $c = 3.24$

19. $A = 78°$, $b = 200$

20. $a = 30$, $b = 82.42$

An observer for a radar station is located at the origin of a coordinate system. For each point, find the bearing of an airplane located at that point. Express the bearing using both methods.

1. (0,4)

2. (3,3)

3. (5,0)

4. (2,–2)

Work each problem. In these exercises, assume the course of a plane or ship is on the indicated bearing.

5. An airplane flying at 650 miles per hour has a bearing of N 42° W. After flying 2.25 hours, how far north and how far west has the plane traveled from its point of departure?

6. A ship leaves port at 8am and has a bearing of S 36° E. If the ship is sailing at 17 knots, how many nautical miles south and how many nautical miles east has the ship traveled by noon?

7. A ship is 35 miles west and 10 miles north of port. If the crew wants to sail directly to port, what bearing should they take?

8. A plane is 300 miles south and 55 miles west of the airport. If the pilot wants to fly directly to the airport, what bearing should he take?

9. A boat leaves a harbor and travels 31 miles in the direction N 36° E. The captain then turns the boat 90° and travels another 27 miles in the direction S 42° E. At that time, how far is the boat from the harbor?

10. Refer back to exercise 9. What is the bearing of the boat from the harbor?

11. A rowboat leaves a dock and travels 3.2 miles in the direction S 29° W. The rower turns the boat 90° and rows 4.6 miles in the direction N 61° W. At that time, how far is the rowboat from the dock?

12. Refer back to exercise 11. What is the rowboat's bearing from the dock?

13. A boat leaves port on a bearing of N 46° W for 92 miles. How many miles north and how many miles west has the ship traveled?

14. A boat leaves port on a bearing of S 54° E for 82 miles. How many miles south and how many miles east has the ship traveled?

15. A boat leaves port on a bearing of N 47° W for 84 miles. How many miles north and how many miles west has the ship traveled?

2.1 Trigonometric Functions of Acute Angles

In exercises 1-2 we give, in order, sine, cosine, tangent, cotangent, secant, and cosecant.

1. $b=5;\ \frac{3}{5};\ \frac{4}{5};\ \frac{3}{4};\ \frac{4}{3};\ \frac{5}{4};\ \frac{5}{3}$

2. $c=17;\ \frac{15}{17};\ \frac{8}{17};\ \frac{15}{8};\ \frac{8}{15};\ \frac{17}{8};\ \frac{17}{15}$

3. $\cos 74°$

4. $\cot 41.73°$

5. $\sin(15° - \alpha)$

6. $\cos(109° - \theta)$

7. 40°

8. 28°

9. 3°

10. 12°

11. False

12. True

13. True

14. False

15. $\frac{\sqrt{2}}{2}$

16. 1

17. $\frac{\sqrt{3}}{2}$

18. $\frac{2\sqrt{3}}{3}$

19. 2

20. $\frac{\sqrt{3}}{3}$

2.2 Trigonometric Functions of Non-Acute Angles

In exercises 1-8 we give, in order, sine, cosine, tangent, cotangent, secant, and cosecant.

1. $\frac{\sqrt{2}}{2};\ -\frac{\sqrt{2}}{2};\ -1;\ -1;\ -\sqrt{2};\ \sqrt{2}$

2. $\frac{1}{2};\ \frac{\sqrt{3}}{2};\ \frac{\sqrt{3}}{3};\ \sqrt{3};\ \frac{2\sqrt{3}}{3};\ 2$

3. $\frac{1}{2};\ \frac{\sqrt{3}}{2};\ \frac{\sqrt{3}}{3};\ \sqrt{3};\ \frac{2\sqrt{3}}{3};\ 2$

4. $-\frac{\sqrt{3}}{2};\ -\frac{1}{2};\ \sqrt{3};\ \frac{\sqrt{3}}{3};\ -2;\ -\frac{2\sqrt{3}}{3}$

5. $-\frac{\sqrt{2}}{2};\ -\frac{\sqrt{2}}{2};\ 1;\ 1;\ -\sqrt{2};\ -\sqrt{2}$

6. $-\frac{\sqrt{3}}{2};\ \frac{1}{2};\ -\sqrt{3};\ -\frac{\sqrt{3}}{3};\ 2;\ -\frac{2\sqrt{3}}{3}$

7. $-\frac{\sqrt{3}}{2};\ \frac{1}{2};\ -\sqrt{3};\ -\frac{\sqrt{3}}{3};\ 2;\ -\frac{2\sqrt{3}}{3}$

8. $\frac{1}{2};\ -\frac{\sqrt{3}}{2};\ -\frac{\sqrt{3}}{3};\ -\sqrt{3};\ -\frac{2\sqrt{3}}{3};\ 2$

9. $\frac{\sqrt{3}}{2}$

10. $-\frac{\sqrt{3}}{2}$

11. $\frac{\sqrt{3}}{3}$

12. True

13. False

14. False

15. Positive

16. Negative

17. Negative

18. 30°, 150°

19. 240°, 300°

20. 30°, 210°

2.3 Finding Trigonometric Function Values Using a Calculator

1. 0.0805468609 **2.** 0.9829674397 **3.** 0.4452286853 **4.** 22.36626841

5. 1.983474007 **6.** −10.92876842 **7.** 0.2915973475 **8.** 0.901564217

9. −1.101461223 **10.** 1.019804143 **11.** 28° **12.** 26.77°

13. 36.3° **14.** 60° **15.** 1 **16.** −1

17. True **18.** False **19.** False **20.** True

2.4 Solving Right Triangles

1. $B = 60°$, $a = 57.7$, $c = 115.5$

2. $A = 55°$, $a = 7.1$, $c = 8.7$

3. $B = 57.4°$, $b = 65$, $c = 77.2$

4. $B = 40°$, $a = 9.2$, $b = 7.7$

5. $A = 50°$, $a = 7$, $b = 5.9$

6. $A = 57.46°$, $b = 4.2$, $c = 7.9$

7. $A = 38°$, $b = 39.7$, $c = 50.4$

8. $B = 15°$, $a = 19.3$, $b = 5.2$

9. $B = 22°$, $a = 198.1$, $c = 213.6$

10. $A = 66.2°$, $B = 23.8°$, $a = 47.79$

11. $A = 60°$, $b = 23.1$, $c = 46.2$

12. $A = 36.9°$, $B = 53.1°$, $c = 25$

13. $A = 52°$, $b = 11.7$, $c = 19$

14. $A = 56.2°$, $B = 33.8°$, $b = 3.661$

15. $A = 31°$, $B = 59°$, $a = 72.1$

16. $A = 33.6°$, $B = 56.4°$, $b = 42.7$

17. $B = 30°$, $a = 19.1$, $b = 10.9$

18. $A = 64.6°$, $a = 2.9$, $b = 1.4$

19. $B = 12°$, $a = 940.9$, $c = 961.9$

20. $A = 20°$, $B = 70°$, $c = 87.71$

2.5 Further Applications of Right Triangles

1. 0°; N 0° E

2. 45°; N 45° E

3. 90°; N 90° E

4. 135°; S 45° E

5. West 978.6 miles, North 1086.8 miles

6. East 40 miles, South 55 miles

7. S 74.05° E

8. N 10.39° E

9. 41.1 miles

10. N 77.1° E

11. 5.6 miles

12. S 84.2° W

13. West 66.2 miles, North 63.9 miles

14. East 66.3 miles, South 48.2 miles

15. West 61.4 miles, North 57.3 miles

Convert each degree measure to radians. Leave answers as multiples of π.

1. $15°$

2. $-328°$

3. $720°$

Convert each radian measure to degrees.

4. $\frac{11\pi}{18}$

5. $\frac{7\pi}{3}$

6. $-\frac{\pi}{80}$

7. $\frac{5\pi}{6}$

Convert each degree measure to radians.

8. $419°$

9. $4.8°$

10. $\pi°$

11. $29°45'$

Convert each radian measure to degrees. Write answers to the nearest minute.

12. 5.9

13. -0.5

14. 2.6

Find the exact value of each expression without using a calculator.

15. $\sin\frac{\pi}{3}$

16. $\cos\frac{\pi}{4}$

17. $\tan\frac{\pi}{6}$

18. $\csc\frac{5\pi}{6}$

19. $\sec\frac{\pi}{2}$

20. $\cot\ -\frac{7\pi}{6}$

Find the length of each arc intercepted by a central angle θ in a circle of radius *r*.

1. $r = 4$ cm; $\theta = \frac{\pi}{8}$

2. $r = 39$ in; $\theta = 25°$

3. $r = 13$ m; $\theta = 21°$

4. $r = 3$ cm; $\theta = 1$

5. $r = 15$ m; $\theta = 76°$

6. $r = 80$ in; $\theta = \frac{3\pi}{2}$

7. $r = 8.4$ m; $\theta = 135°$

Find the area of a sector of a circle having radius *r* and central angle θ.

8. $r = 1.4$ ft; $\theta = \frac{5\pi}{4}$

9. $r = 5$ in; $\theta = 2$

10. $r = 4$ cm; $\theta = 3\pi$

11. $r = 15$ ft; $\theta = 31°$

12. $r = 9.5$ in; $\theta = 24°$

13. $r = 9$ cm; $\theta = 60°$

14. $r = 4$ ft; $\theta = \frac{7\pi}{3}$

15. $r = 12$ cm; $\theta = \frac{2\pi}{3}$

16. $r = 3$ in; $\theta = \frac{10\pi}{11}$

17. $r = 4$ in; $\theta = \frac{\pi}{2}$

18. $r = 13.2$ ft; $\theta = 34°$

19. $r = 0.76$ cm; $\theta = 60°$

20. $r = 6$ m; $\theta = 4.5$

For each value of the real number *s*, find (a) sin *s*, (b) cos *s*, and (c) tan *s*.

1. $\frac{3\pi}{2}$

2. 4π

3. $-\frac{\pi}{2}$

Find the exact circular function value for each of the following.

4. $\csc\frac{5\pi}{6}$

5. $\tan\frac{11\pi}{6}$

6. $\sin\frac{13\pi}{6}$

7. $\tan 12\pi$

8. $\sin -\frac{8\pi}{3}$

9. $\sin -\frac{5\pi}{3}$

10. $\tan \frac{7\pi}{4}$

Find a calculator approximation for each circular function value.

11. sec(1.324)

12. cos(–4)

13. sin(2.268)

14. csc(2.805)

15. cot(–2.512)

Find the value of *s* in the interval $[0,\frac{\pi}{2}]$ that makes each statement true.

16. $\sin s = 0.7146$

17. $\cos s = 0.1472$

18. $\tan s = 3.3694$

Find the exact value of *s* in the given interval that has the given circular function value. Do not use a calculator,

19. $[-\frac{\pi}{2},-\pi]$; $\sin s = -\frac{1}{2}$

20. $[\pi,\frac{3\pi}{2}]$; $\tan s = 1$

21. $[\frac{\pi}{2},\pi]$; $\sin s = \frac{\sqrt{3}}{2}$

22. $[\frac{3\pi}{2},2\pi]$; $\cos s = \frac{1}{2}$

23. $[\frac{\pi}{2},\pi]$; $\cos s = -\frac{\sqrt{3}}{2}$

Use the formula $\omega = \theta / t$ to find the value of the missing variable.

1. $\theta = \frac{3\pi}{4}$; $t = 2$ min

2. $\omega = 5$ rad/sec; $t = 3$ sec

3. $\theta = 12\pi$; $\omega = 10.47$ rad/hr

4. $\omega = 4$ rad/sec; $t = 10$ sec

5. $\theta = 12$; $t = 3$ min

6. $\omega = \frac{4\pi}{3}$ rad/sec; $\theta = 40\pi$

7. $\theta = 15\pi$; $t = 2.4$ hr

8. $\omega = \frac{4\pi}{5}$ rad/sec; $t = 6$ sec

Use the formula $v = r \cdot \omega$ to find the value of the missing variable.

9. $\omega = \frac{\pi}{12}$ rad/sec; $r = 936$ ft

10. $\omega = 8$ rad/sec; $r = 4$ in

11. $r = 3$ cm; $v = 6$ cm/sec

12. $r = 30$ ft; $v = 9$ ft/min

13. $v = 31$ ft/sec; $\omega = 6.5$ rad/sec

14. $\omega = 9.2$ rad/sec; $r = 4$ ft

Use the formula $s = r \cdot \omega \cdot t$ to find the value of the missing variable.

15. $s = 17.3$ in; $r = 3$ in; $\omega = 1.5$ rad/sec

16. $s = 9.5$ cm; $r = 35$ cm; $t = 2$ hr

17. $s = 23.6$ m; $\omega = 3$ rad/min; $t = 6$ min

18. $r = 3.7$ ft; $\omega = 7.5$ rad/s; $t = 10$ sec

19. $s = 100$ ft; $r = 4$ ft; $\omega = 11$ rad/min

20. $s = 2.17$ mi; $r = 10$ mi; $t = 5$ sec

3.1 Radian Measure

1. $\frac{\pi}{12}$

2. $-\frac{82\pi}{45}$

3. 4π

4. 110°

5. 420°

6. −2.25°

7. 150°

8. 7.31293

9. 0.083776

10. 0.054831

11. 0.519235

12. 338°3′

13. −28°39′

14. 148°58′

15. $\frac{\sqrt{3}}{2}$

16. $\frac{\sqrt{2}}{2}$

17. $\frac{\sqrt{3}}{3}$

18. 2

19. undefined

20. $-\frac{\sqrt{3}}{3}$

3.2 Applications of Radian Measure

1. 1.57 cm

2. 17.02 in

3. 4.76 m

4. 3 cm

5. 19.90 m

6. 376.99 in

7. 19.79 m

8. 3.85 ft^2

9. 25 in^2

10. 75.4 cm^2

11. 60.87 ft^2

12. 18.9 in^2

13. 42.41 cm^2

14. 58.64 ft^2

15. 150.80 cm^2

16. 12.85 in^2

17. 12.57 in^2

18. 51.698 ft^2

19. 0.30 cm^2

20. 81 m^2

3.3 The Unit Circle and Circular Functions

1. (a) −1, (b) 0, (c) undefined **2.** (a) 0, (b) 1, (c) 0 **3.** (a) −1, (b) 0, (c) undefined

4. 2 **5.** $-\frac{\sqrt{3}}{3}$ **6.** $\frac{1}{2}$ **7.** 0

8. $-\frac{\sqrt{3}}{2}$ **9.** $\frac{\sqrt{3}}{2}$ **10.** −1 **11.** 4.093

12. −0.654 **13.** 0.767 **14.** 3.028 **15.** 1.373

16. 0.7961 **17.** 1.4231 **18.** 1.2823 **19.** $-\frac{5\pi}{6}$

20. $\frac{5\pi}{4}$ **21.** $\frac{2\pi}{3}$ **22.** $\frac{5\pi}{3}$ **23.** $\frac{5\pi}{6}$

3.4 Linear and Angular Speed

1. $\frac{3\pi}{8}$ rad/min **2.** 15 radians **3.** 3.6 hours **4.** 40 radians

5. 4 rad/min **6.** 30 seconds **7.** 19.64 rad/hr **8.** $\frac{24\pi}{5}$ radians

9. 78 ft/s **10.** 32 in/s **11.** 2 rad/sec **12.** $\frac{3}{10}$ rad/min

13. 4.77 feet **14.** 36.8 ft/sec **15.** 3.84 seconds **16.** 0.14 rad/hr

17. 1.31 meters **18.** 277.5 feet **19.** 2.27 minutes **20.** 0.0434 rad/s

Find the amplitude of the graph of each function. Then give a calculator window that shows the function over a two-period interval.

1. $y = 5\cos x$

2. $y = -4\sin x$

3. $y = 7\cos x$

4. $y = 2\sin x$

5. $y = -4\cos x$

6. $y = \frac{5}{8}\sin x$

7. $y = 8\cos x$

Find the amplitude and the period of the graph of each function. Then give a calculator window that shows the function over a two-period interval.

8. $y = -\frac{1}{2}\cos(\pi x)$

9. $y = \sin(2x)$

10. $y = \frac{1}{4}\sin(\pi x)$

11. $y = -3\cos\frac{x}{4}$

12. $y = 3\sin(4\pi x)$

13. $y = -2\sin(6x)$

14. $y = -3\sin\frac{x}{2}$

15. $y = -\frac{4}{3}\cos(3x)$

16. $y = \frac{1}{2}\sin\frac{\pi x}{3}$

17. $y = \cos\frac{1}{3}x$

18. $y = 4\sin(6x)$

19. $y = 3\cos\frac{x}{2}$

20. $y = -\frac{4}{3}\sin\frac{x}{\pi}$

Find the amplitude, the period, any vertical translation and any phase shift of the graph of each function. Then give a calculator window that shows the function over a two-period interval.

1. $y = \cos\ x - \dfrac{\pi}{2}$

2. $y = \cos\ x - \dfrac{\pi}{4}$

3. $y = 4\cos\ x + \dfrac{\pi}{4}\ + 5$

4. $y = \dfrac{1}{2}\sin\ \dfrac{\pi x}{2}$

5. $y = -2 + 4\cos(2x)$

6. $y = -2 + 2\cos\ \dfrac{x}{2} + \dfrac{\pi}{2}$

7. $y = 3 + 3\sin\ \dfrac{x}{2} + \dfrac{\pi}{3}$

8. $y = -4\cos\ \dfrac{2x}{3} + \dfrac{\pi}{6}$

9. $y = 6\sin\ \dfrac{x}{3} - \dfrac{\pi}{6}$

10. $y = 4\sin(\pi x - 2) + 1$

11. $y = \sin\ x + \dfrac{\pi}{6}$

12. $y = 2\sin\ 3x - \dfrac{\pi}{2}$

13. $y = \cos(2x - \pi)$

14. $y = 3\sin(2\pi x) - 2$

15. $y = -\sin\ x - \dfrac{\pi}{2}$

16. $y = 2\cos\ x - \dfrac{5\pi}{6}$

17. $y = \dfrac{1}{2}\sin\ x + \dfrac{\pi}{4}$

18. $y = 4\sin\ \dfrac{x}{2} + \dfrac{\pi}{2}$

19. $y = 2\sin\ \dfrac{x}{3}\ - 1$

20. $y = \cos\ \dfrac{x}{3} - \dfrac{\pi}{2}\ - 1$

Graph each function over a one-period interval. Give a calculator window that shows the graph.

1. $y=3\tan\ \frac{x}{3}$

2. $y=\tan\ x-\frac{\pi}{4}$

3. $y=-2\tan(3x)$

4. $y=2\cot\ 2x-\frac{\pi}{4}$

5. $y=-\cot\ \frac{x}{\pi}$

6. $y=-\cot(4x-\pi)$

7. $y=\cot\ x+\frac{\pi}{4}$

8. $y=\tan(2x+\pi)$

9. $y=-3\cot\ \frac{x}{4}+2\pi$

10. $y=\cot(3x)$

11. $y=\tan(\pi x)-1$

12. $y=3\tan\ 2x-\frac{\pi}{3}$

13. $y=\tan\ \frac{x}{2}\ -4$

14. $y=3\tan\ 2x+\frac{\pi}{3}$

15. $y=\tan\ x+\frac{\pi}{4}$

16. $y=3\tan\ \frac{\pi x}{2}$

17. $y=\cot\ 2x+\frac{\pi}{2}$

18. $y=\cot\ \frac{x}{2}\ -1$

19. $y=-2\cot(4x)+2$

20. $y=-2+\cot\ \frac{x}{3}-\frac{\pi}{2}$

Graph each function over a one-period interval. Give a calculator window that shows the graph.

1. $y = 3\sec \frac{x}{3}$

2. $y = \sec x - \frac{\pi}{4}$

3. $y = -2\sec(3x)$

4. $y = 2\csc 2x - \frac{\pi}{4}$

5. $y = -\csc \frac{x}{\pi}$

6. $y = -\csc(4x - \pi)$

7. $y = \csc x + \frac{\pi}{4}$

8. $y = \sec(2x + \pi)$

9. $y = -3\csc \frac{x}{4} + 2\pi$

10. $y = \csc(3x)$

11. $y = \sec(\pi x) - 1$

12. $y = 3\sec 2x - \frac{\pi}{3}$

13. $y = \sec \frac{x}{2} - 4$

14. $y = 3\sec 2x + \frac{\pi}{3}$

15. $y = \sec x + \frac{\pi}{4}$

16. $y = 3\sec \frac{\pi x}{2}$

17. $y = \csc 2x + \frac{\pi}{2}$

18. $y = \csc \frac{x}{2} - 1$

19. $y = -2\csc(4x) + 2$

20. $y = -2 + \csc \frac{x}{3} - \frac{\pi}{2}$

Suppose that a weight on a spring has initial position $s(0)$ and period P. Find a function s given by $s(t) = a\cos(\omega t)$ that models the displacement of the weight.

1. $s(0) = 3$ inches; $P = 0.6$ second

2. $s(0) = 6$ inches; $P = 2.3$ seconds

3. $s(0) = -2$ inches; $P = 0.4$ second

4. $s(0) = -5$ inches; $P = 1.5$ seconds

A note on the piano has given frequency F. Suppose the maximum displacement at the center of the piano wire is given by $s(0)$. Find constants a and ω so that the equation $s(t) = a\cos(\omega t)$ models this displacement.

5. $F = 32$; $s(0) = 0.17$

6. $F = 140$; $s(0) = 0.07$

7. $F = 66$; $s(0) = 0.24$

8. $F = 280$; $s(0) = 0.12$

The period P, in seconds, of oscillation of a pendulum of length L feet is given by $P = 2\pi\sqrt{\frac{L}{32}}$. Find the missing value.

9. $L = 1$ foot

10. $L = 2.5$ feet

11. $P = 6$ seconds

12. $P = 0.5$ second

Solve each problem.

13. The equation for the simple harmonic motion of a particle is given by $s(t) = 3\cos\frac{3\pi}{2}t$. Find the maximum displacement and the frequency.

14. The position of a weight attached to a spring is $s(t) = -8\cos(16\pi t)$ inches after t seconds. What is the maximum height that the weight rises above the equilibrium position and when does the weight first reach its maximum height?

15. The position of a weight attached to a spring is $s(t) = -\frac{1}{2}\cos(20\pi t)$ inches after t seconds. What are the frequency and period?

16. The position of a weight attached to a spring is $s(t) = -8\cos(60\pi t)$ inches after t seconds. Calculate and interpret $s(0.26)$.

17. A weight attached to a spring is pulled down 6 inches below the equilibrium position. Assuming that the frequency is $7/\pi$ cycles per second, determine a trigonometric model that gives the position of the weight at time t seconds.

4.1 Graphs of the Sine and Cosine Functions

1. 5; $[-2\pi,2\pi]$ by $[-6,6]$

2. 4; $[-2\pi,2\pi]$ by $[-5,5]$

3. 7; $[-2\pi,2\pi]$ by $[-8,8]$

4. 2; $[-2\pi,2\pi]$ by $[-3,3]$

5. 4; $[-2\pi,2\pi]$ by $[-5,5]$

6. $\frac{5}{8}$; $[-2\pi,2\pi]$ by $[-1,1]$

7. 8; $[-2\pi,2\pi]$ by $[-9,9]$

8. $\frac{1}{2}$; 2; $[-2,2]$ by $[-1,1]$

9. 1; π; $[-\pi,\pi]$ by $[-2,2]$

10. $\frac{1}{4}$; 2; $[-2,2]$ by $[-1,1]$

11. 3; 8π; $[-8\pi,8\pi]$ by $[-4,4]$

12. 3; $\frac{1}{2}$; $[-\frac{1}{2},\frac{1}{2}]$ by $[-4,4]$

13. 2; $\frac{\pi}{3}$; $[-\frac{\pi}{3},\frac{\pi}{3}]$ by $[-3,3]$

14. 3; 4π; $[-4\pi,4\pi]$ by $[-4,4]$

15. $-\frac{4}{3}$; $\frac{2\pi}{3}$; $[-\frac{2\pi}{3},\frac{2\pi}{3}]$ by $[-2,2]$

16. $\frac{1}{2}$; 6; $[-6,6]$ by $[-1,1]$

17. 1; 6π; $[-6\pi,6\pi]$ by $[-2,2]$

18. 4; $\frac{\pi}{3}$; $[-\frac{\pi}{3},\frac{\pi}{3}]$ by $[-5,5]$

19. 3; 4π; $[-4\pi,4\pi]$ by $[-4,4]$

20. $\frac{4}{3}$; $2\pi^2$; $[-2\pi^2,2\pi^2]$ by $[-2,2]$

4.2 Translations of the Graphs of the Sine and Cosine Functions

1. 1; 2π; none; $\frac{\pi}{2}$ right; $[-\frac{3\pi}{2},\frac{5\pi}{2}]$ by $[-2,2]$

2. 1; 2π; none; $\frac{\pi}{4}$ right; $[-\frac{7\pi}{4},\frac{9\pi}{4}]$ by $[-2,2]$

3. 4; 2π; 5 up; $\frac{\pi}{4}$ left; $[-\frac{9\pi}{4},\frac{7\pi}{4}]$ by $[0,10]$

4. $\frac{1}{2}$; 4; none; none; $[-4,4]$ by $[-1,1]$

5. 4; π; 2 down; none; $[-\pi,\pi]$ by $[-7,3]$

6. 2; 4π; 2 down; π left; $[-5\pi,3\pi]$ by $[-4,0]$

7. 3; 4π; 3 up; $\frac{2\pi}{3}$ left; $[-\frac{14\pi}{3},\frac{10\pi}{3}]$ by $[0,6]$

8. 4; 3π; none; $\frac{\pi}{4}$ left; $[-\frac{13\pi}{4},\frac{11\pi}{4}]$ by $[-5,5]$

9. 6; 6π; none; $\frac{\pi}{2}$ right; $[-\frac{11\pi}{2},\frac{13\pi}{2}]$ by $[-7,7]$

10. 4; 2; 1 up; $\frac{2}{\pi}$ right; $[-2+\frac{2}{\pi},2+\frac{2}{\pi}]$ by $[-4,6]$

11. 1; 2π; none; $\frac{\pi}{6}$ left; $[-\frac{13\pi}{6},\frac{11\pi}{6}]$ by $[-2,2]$

12. 2; $\frac{2\pi}{3}$; none; $\frac{\pi}{6}$ right; $[-\frac{\pi}{2},\frac{5\pi}{6}]$ by $[-3,3]$

13. 1; π; none; $\frac{\pi}{2}$ right; $[-\frac{\pi}{2},\frac{3\pi}{2}]$ by $[-2,2]$

14. 3; 1; 2 down; none; $[-1,1]$ by $[-6,2]$

15. 1; 2π; none; $\frac{\pi}{2}$ right; $[-\frac{3\pi}{2},\frac{5\pi}{2}]$ by $[-2,2]$

16. 2; 2π; none; $\frac{5\pi}{6}$ right; $[-\frac{7\pi}{6},\frac{17\pi}{6}]$ by $[-3,3]$

17. $\frac{1}{2}$; 2π; none; $\frac{\pi}{4}$ left; $[-\frac{9\pi}{4},\frac{7\pi}{4}]$ by $[-1,1]$

18. 4; 4π; none; π left; $[-5\pi,3\pi]$ by $[-5,5]$

19. 2; 6π; 1 down; none; $[-6\pi,6\pi]$ by $[-4,2]$

20. 1; 6π; 1 down; $\frac{3\pi}{2}$ right; $[-\frac{9\pi}{2},\frac{15\pi}{2}]$ by $[-3,1]$

4.3 Graphs of the Tangent and Cotangent Functions

1. $[-\frac{3\pi}{2},\frac{3\pi}{2}]$ by $[-10,10]$

2. $[4-\frac{\pi}{2},4+\frac{\pi}{2}]$ by $[-10,10]$

3. $[-\frac{\pi}{6},\frac{\pi}{6}]$ by $[-10,10]$

4. $[\frac{\pi}{8},\frac{5\pi}{8}]$ by $[-10,10]$

5. $[0,\pi^2]$ by $[-10,10]$

6. $[\frac{\pi}{4},\frac{\pi}{2}]$ by $[-10,10]$

7. $[-\frac{\pi}{4},\frac{3\pi}{4}]$ by $[-10,10]$

8. $[-\frac{3\pi}{4},-\frac{\pi}{4}]$ by $[-10,10]$

9. $[0,4\pi]$ by $[-10,10]$

10. $[0,\frac{\pi}{3}]$ by $[-10,10]$

11. $[-\frac{1}{2},\frac{1}{2}]$ by $[-10,10]$

12. $[-\frac{\pi}{12},\frac{5\pi}{12}]$ by $[-10,10]$

13. $[-\pi,\pi]$ by $[-10,10]$

14. $[-\frac{5\pi}{12},\frac{\pi}{12}]$ by $[-10,10]$

15. $[-\frac{3\pi}{4},\frac{\pi}{4}]$ by $[-10,10]$

16. $[-1,1]$ by $[-10,10]$

17. $[-\frac{\pi}{4},\frac{\pi}{4}]$ by $[-10,10]$

18. $[0,2\pi]$ by $[-10,10]$

19. $[0, \frac{\pi}{4}]$ by $[-10,10]$

20. $[-\frac{3\pi}{2}, \frac{3\pi}{2}]$ by $[-10,10]$

4.4 Graphs of the Secant and Cosecant Functions

1. $[0, 6\pi]$ by $[-10,10]$

2. $[\frac{\pi}{4}, \frac{9\pi}{4}]$ by $[-10,10]$

3. $[0, \frac{2\pi}{3}]$ by $[-10,10]$

4. $[\frac{\pi}{8}, \frac{9\pi}{8}]$ by $[-10,10]$

5. $[0, 2\pi^2]$ by $[-10,10]$

6. $[\frac{\pi}{4}, \frac{3\pi}{4}]$ by $[-10,10]$

7. $[-\frac{\pi}{4}, \frac{7\pi}{4}]$ by $[-10,10]$

8. $[-\frac{\pi}{2}, \frac{\pi}{2}]$ by $[-10,10]$

9. $[0, 8\pi]$ by $[-10,10]$

10. $[0, \frac{2\pi}{3}]$ by $[-10,10]$

11. $[0,2]$ by $[-10,10]$

12. $[\frac{\pi}{6}, \frac{7\pi}{6}]$ by $[-10,10]$

13. $[0, 4\pi]$ by $[-10,10]$

14. $[-\frac{\pi}{6}, \frac{5\pi}{6}]$ by $[-10,10]$

15. $[-\frac{\pi}{4}, \frac{7\pi}{4}]$ by $[-10,10]$

16. $[0,4]$ by $[-10,10]$

17. $[-\frac{\pi}{4}, \frac{3\pi}{4}]$ by $[-10,10]$

18. $[0, 4\pi]$ by $[-10,10]$

19. $[0, \frac{\pi}{2}]$ by $[-10,10]$

20. $[\frac{3\pi}{2}, \frac{15\pi}{2}]$ by $[-10,10]$

4.5 Harmonic Motion

1. $s(t) = 3\cos \frac{10\pi}{3} t$

2. $s(t) = 6\cos \frac{20\pi}{23} t$

3. $s(t) = -2\cos(5\pi t)$

4. $s(t) = -5\cos \frac{4\pi}{3} t$

5. $s(t) = 0.17\cos(64\pi t)$

6. $s(t) = 0.07\cos(280\pi t)$

7. $s(t) = 0.24\cos(132\pi t)$

8. $s(t) = 0.12\cos(560\pi t)$

9. $\sqrt{2}\pi/4$ sec

10. $\sqrt{5}\pi/4$ sec

11. $288/\pi^2$ feet

12. $2/\pi^2$ feet

13. 3; 3/4

14. 8; 1/16 sec

15. 10 cycles/sec; 1/10 sec

16. -2.47; After 0.26 seconds the weight is about 2.5 inches below the equilibrium position.

17. $s(t) = -6\cos(14t)$

Find sin θ.

1. $\cos\theta = -\frac{12}{13}$, θ in quadrant III

2. $\tan\theta = \frac{9}{40}$, θ in quadrant I

3. $\sec\theta = -\frac{4}{\sqrt{5}}$, $\tan\theta < 0$

Find the remaining five trigonometric functions of θ.

4. $\sin\theta = -\frac{8}{17}$, $\cos\theta < 0$

5. $\tan\theta = \frac{1}{2}$, θ in quadrant I

6. $\sec\theta = 3$, θ in quadrant IV

7. $\sin\theta = \frac{1}{3}$, θ in quadrant II

8. $\csc\theta = \sqrt{3}$, $\cos\theta > 0$

Write each expression in terms of sine and cosine, and simplify so that no quotients appear in the final expression.

9. $\tan^2\theta \cdot \csc\theta \cdot \cos\theta$

10. $\sin\theta \cdot \cos\theta \cdot \csc\theta$

11. $(\sin\theta + \cos\theta)^2 - 2\sin\theta \cdot \cos\theta$

12. $\tan^2\theta - \cot^2\theta$

13. $\sin^2\theta + \cot^2\theta \cdot \sin^2\theta$

14. $(1 + \sin\theta)(1 - \sin\theta)$

15. $\frac{1}{\sin^2\theta} - \frac{1}{\tan^2\theta}$

16. $\tan\theta(\cos\theta - \csc\theta)$

17. $\sin\theta(\sec\theta + \cot\theta)$

Use a graphing calculator to make a conjecture as to whether each equation is an identity.

18. $\tan^2 x = \sec^2 x - 1$

19. $\cos^2 x = 2 + \cos 2x$

20. $\tan 2x = 1 - \tan^2 x$

Perform each indicated operation and simplify the result.

1. $\dfrac{1}{\sec x+1}-\dfrac{1}{\sec x-1}$

2. $(1-\csc x)(1+\csc x)$

3. $\dfrac{\cos x+\sin x}{\cos x}$

4. $\cos x(\csc x-\sec x)$

Factor each trigonometric expression.

5. $\tan^2\theta-\tan^2\theta\sin^2\theta$

6. $\sec^2 x-4$

7. $\cos^4\theta-\sin^4\theta$

Use fundamental identities to simplify each expression.

8. $\tan\theta\csc\theta$

9. $\sin^2\theta(\cot^2\theta+1)$

10. $\dfrac{\csc\theta}{\sec\theta}$

Verify that each trigonometric equation is an identity.

11. $\dfrac{\cot^3\theta}{\csc\theta}=\cos\theta(\csc^2\theta-1)$

12. $\dfrac{\tan x\cdot\cos x}{\sin x}=1$

13. $\sec x-\cos x=\tan x\cdot\sin x$

14. $\dfrac{\sin x+\cos x}{\sin x}=1+\cot x$

15. $\dfrac{\sin x-\cos x}{\sin x\cdot\cos x}=\sec x-\csc x$

16. $\dfrac{1}{\sec x\cdot\tan x}=\csc x-\sin x$

17. $\tan x(1-\cot x)=\tan x-1$

18. $(\sin x+\cos x)^2+(\sin x-\cos x)^2=2$

19. $\sec^2 x+\sec x\cdot\tan x=\dfrac{1}{1-\sin x}$

20. $\dfrac{1+\cos x}{\sin x}=\dfrac{1}{\csc x-\cot x}$

Use identities to find each exact value.

1. $\cos 105°$

2. $\cos 255°$

3. $\cos \dfrac{11\pi}{12}$

4. $\cos \dfrac{10\pi}{9} \cdot \cos \dfrac{5\pi}{36} - \sin \dfrac{10\pi}{9} \cdot \sin \dfrac{5\pi}{36}$

Write each function value in terms of the cofunction of a complementary angle.

5. $\sin 75°$

6. $\cos 34°$

7. $\tan \dfrac{\pi}{10}$

8. $\cos 75°$

9. $\sin \dfrac{\pi}{12}$

Find an angle θ that makes each statement true.

10. $\sin\theta = \cos(\theta + 10°)$

11. $\tan(2\theta - 14°) = \cot(\theta + 20°)$

12. $\sec\theta = \csc(\theta - 30°)$

Use identities to write each expression as a function of x or θ.

13. $\cos(x + 180°)$

14. $\cos(2\pi - \theta)$

15. $\cos(\dfrac{3\pi}{2} - \theta)$

Use the given information to find $\cos(s+t)$.

16. $\sin s = \dfrac{5}{13}$; $\cos t = -\dfrac{3}{5}$; s in quadrant I, t in quadrant II

17. $\sin s = \dfrac{3}{5}$; $\cos t = -\dfrac{5}{13}$; s in quadrant I, t in quadrant II

18. $\sin s = \dfrac{7}{25}$; $\cos t = -\dfrac{8}{17}$; s in quadrant II, t in quadrant III

Verify that each equation is an identity.

19. $\cos(x+y)\cos(x-y) = \cos^2 x - \sin^2 y$

20. $\cos(x+y) + \cos(x-y) = 2\cos x \cos y$

Use identities to find each exact value.

1. $\sin \frac{21\pi}{12}$
2. $\sin(165°)$
3. $\sin \frac{7\pi}{12}$
4. $\tan(105°)$
5. $\tan(-\frac{\pi}{12})$
6. $\tan(-195°)$

Use identities to write each expression as a function of x or θ.

7. $\sin(\frac{3\pi}{2}+\theta)$
8. $\tan(\frac{\pi}{4}-x)$
9. $\sin(\theta+\frac{\pi}{2})$
10. $\tan(\theta+\frac{\pi}{4})$
11. $\sin(x-\pi)$
12. $\tan(\pi+x)$

Use the given information to find $\sin(s+t)$ and $\tan(s-t)$.

13. $\sin s=\frac{5}{13}$; $\cos t=-\frac{3}{5}$; s in quadrant I, t in quadrant II

14. $\sin s=\frac{3}{5}$; $\cos t=-\frac{5}{13}$; s in quadrant I, t in quadrant II

15. $\sin s=\frac{7}{25}$; $\cos t=-\frac{8}{17}$; s in quadrant II, t in quadrant III

Find each exact value.

16. $\sin(105°)$
17. $\tan \frac{17\pi}{12}$
18. $\sin \frac{13\pi}{12}$

Verify that each equation is an identity.

19. $\sin(x+y)\cdot\sin(x-y)=\sin^2 x-\sin^2 y$
20. $\frac{\cos(x+y)}{\sin x\cdot\cos y}=\cot x-\tan y$

Use identities to find values of the sine and cosine functions for each angle measure.

1. θ, given $\cos 2\theta = \frac{5}{13}$ and $0 < \theta < \frac{\pi}{2}$

2. θ, given $\cos 2\theta = \frac{1}{3}$ and $\pi < \theta < \frac{3\pi}{2}$

3. 2θ, given $\tan\theta = \frac{5}{12}$ and $\cos\theta > 0$

4. 2θ, given $\sin\theta = \frac{24}{25}$ and $\tan\theta < 0$

5. 2θ, given $\cos\theta = -\frac{4}{5}$ and $\frac{\pi}{2} < \theta < \pi$

Use an identity to write each expression as a single trigonometric function value or as a single number.

6. $2\sin 30°\cos 30°$

7. $\dfrac{2\tan 75°}{1 - \tan^2 75°}$

8. $\cos^2\frac{\pi}{8} - \sin^2\frac{\pi}{8}$

9. $1 - 2\sin^2 -\frac{\pi}{8}$

10. $1 - 2\sin^2 105°$

11. $\dfrac{2\tan\frac{3\pi}{8}}{1 - \tan^2\frac{3\pi}{8}}$

Verify that each equation is an identity.

12. $\cos 8\theta = \cos^2 4\theta - \sin^2 4\theta$

13. $(\sin\theta + \cos\theta)^2 = 1 + \sin 2\theta$

14. $\dfrac{2\tan\theta}{1 + \tan^2\theta} = \sin 2\theta$

15. $\dfrac{\cos 2\theta}{\sin^2\theta} = \csc^2\theta - 2$

16. $\sin^2\theta = \dfrac{1 - \cos 2\theta}{2}$

17. $\sin^2 x + \cos 2x = \cos^2 x$

18. $\cos 2x = \cos^4 x - \sin^4 x$

19. $\cos 2\theta = \dfrac{1 - \tan^2\theta}{1 + \tan^2\theta}$

20. $1 = 2\sin^2 x + \cos 2x$

Use a half-angle identity to find each exact value.

1. $\tan 15°$

2. $\cos\dfrac{5\pi}{8}$

3. $\sin\dfrac{\pi}{8}$

Find each of the following.

4. $\sin\dfrac{\theta}{2}$, given $\cos\theta=\dfrac{1}{2}$ with $\dfrac{3\pi}{2}<\theta<2\pi$

5. $\cos\dfrac{\theta}{2}$, given $\sin\theta=\dfrac{3}{5}$ with $\dfrac{\pi}{2}<\theta<\pi$

6. $\tan\dfrac{\theta}{2}$, given $\sec\theta=-\dfrac{41}{40}$ with $\dfrac{\pi}{2}<\theta<\pi$

7. $\sin\dfrac{\theta}{2}$, given $\sin\theta=\dfrac{5}{13}$ with $\dfrac{\pi}{2}<\theta<\pi$

8. $\cos\dfrac{\theta}{2}$, given $\cot\theta=3$ with $\pi<\theta<\dfrac{3\pi}{2}$

9. $\tan\dfrac{\theta}{2}$, given $\cos\theta=\dfrac{8}{17}$ with $\dfrac{3\pi}{2}<\theta<2\pi$

Use an identity to write each expression as a single trigonometric function.

10. $\sqrt{\dfrac{1-\cos 30°}{2}}$

11. $\sqrt{\dfrac{1+\cos 80°}{2}}$

12. $\dfrac{\sin 50°}{1+\cos 50°}$

13. $\dfrac{\sin 2\pi}{1+\cos 2\pi}$

14. $\sqrt{\dfrac{1-\cos 200°}{2}}$

15. $\sqrt{\dfrac{1+\cos 550°}{2}}$

Verify that each equation is an identity.

16. $\sin^2\dfrac{\theta}{2}=\dfrac{1}{2}(1-\cos\theta)$

17. $\cos^2\dfrac{\theta}{2}=\dfrac{\tan\theta+\sin\theta}{2\tan\theta}$

18. $\sin^2\dfrac{x}{2}+\cos x=\cos^2\dfrac{x}{2}$

19. $\cos\dfrac{x}{2}-\sin\dfrac{x}{2}\;^{2}=1-\sin x$

20. $\tan\dfrac{\theta}{2}=\csc\theta-\cot\theta$

5.1 Fundamental Identities

1. $-\frac{5}{13}$ **2.** $\frac{9}{41}$ **3.** $\frac{\sqrt{11}}{4}$

4. $\cos\theta = -\frac{15}{17}$, $\tan\theta = \frac{8}{15}$, $\csc\theta = -\frac{17}{8}$, $\sec\theta = -\frac{17}{15}$, $\cot\theta = \frac{15}{8}$

5. $\sin\theta = \frac{\sqrt{5}}{5}$, $\cos\theta = \frac{2\sqrt{5}}{5}$, $\csc\theta = \sqrt{5}$, $\sec\theta = \frac{\sqrt{5}}{2}$, $\cot\theta = 2$

6. $\sin\theta = -\frac{2\sqrt{2}}{3}$, $\cos\theta = \frac{1}{3}$, $\tan\theta = -2\sqrt{2}$, $\csc\theta = -\frac{3\sqrt{2}}{4}$, $\cot\theta = -\frac{\sqrt{2}}{4}$

7. $\cos\theta = -\frac{2\sqrt{2}}{3}$, $\tan\theta = -\frac{\sqrt{2}}{4}$, $\csc\theta = 3$, $\sec\theta = -\frac{3\sqrt{2}}{4}$, $\cot\theta = -2\sqrt{2}$

8. $\sin\theta = \frac{\sqrt{3}}{3}$, $\cos\theta = \frac{\sqrt{6}}{3}$, $\tan\theta = \frac{\sqrt{2}}{2}$, $\sec\theta = \frac{\sqrt{6}}{2}$, $\cot\theta = \sqrt{2}$

9. $\tan\theta$ **10.** $\cos\theta$ **11.** 1

12. $\sec^2\theta - \csc^2\theta$ **13.** 1 **14.** $\cos^2\theta$

15. 1 **16.** $\sin\theta - \sec\theta$ **17.** $\tan\theta + \cos\theta$

18. identity **19.** not an identity **20.** not an identity

5.2 Verifying Trigonometric Identities

1. $-2\cot^2 x$ **2.** $-\cot^2 x$ **3.** $1 + \tan x$

4. $\cot x - 1$ **5.** $\sin^2\theta$ **6.** $(\sec x + 2)(\sec x - 2)$

7. $(\cos\theta - \sin\theta)(\cos\theta + \sin\theta)$ **8.** $\sec\theta$ **9.** 1

10. $\cot\theta$

5.3 Sum and Difference Identities for Cosine

1. $\frac{-\sqrt{6}+\sqrt{2}}{4}$

2. $\frac{-\sqrt{6}+\sqrt{2}}{4}$

3. $\frac{-\sqrt{6}-\sqrt{2}}{4}$

4. $-\frac{\sqrt{2}}{2}$

5. $\cos 15°$

6. $\sin 56°$

7. $\cot \frac{2\pi}{5}$

8. $\sin 15°$

9. $\cos \frac{5\pi}{12}$

10. $40°$

11. $28°$

12. $60°$

13. $-\cos x$

14. $\cos\theta$

15. $-\sin\theta$

16. $-\frac{56}{65}$

17. $-\frac{56}{65}$

18. $\frac{297}{425}$

5.4 Sum and Difference Identities for Sine and Tangent

1. $-\frac{\sqrt{2}}{2}$

2. $\frac{\sqrt{6}-\sqrt{2}}{4}$

3. $\frac{\sqrt{6}+\sqrt{2}}{4}$

4. $-\sqrt{3}-2$

5. $\sqrt{3}-2$

6. $\sqrt{3}-2$

7. $-\cos\theta$

8. $\frac{1-\tan x}{1+\tan x}$

9. $\cos\theta$

10. $\frac{1+\tan\theta}{1-\tan\theta}$

11. $-\sin x$

12. $\tan x$

13. $\frac{33}{65}, \frac{63}{16}$

14. $\frac{33}{65}, -\frac{63}{16}$

15. $\frac{304}{425}, -\frac{416}{87}$

16. $\frac{\sqrt{6}+\sqrt{2}}{4}$

17. $\sqrt{3}+2$

18. $\frac{-\sqrt{6}+\sqrt{2}}{4}$

5.5 Double-Angle Identities

1. $\frac{2\sqrt{13}}{13}; \frac{3\sqrt{13}}{13}$

2. $-\frac{\sqrt{3}}{3}; -\frac{\sqrt{6}}{3}$

3. $\frac{120}{169}; \frac{119}{169}$

4. $-\frac{336}{625}; -\frac{527}{625}$

5. $-\frac{24}{25}; \frac{7}{25}$

6. $\frac{\sqrt{3}}{2}$

7. $-\frac{\sqrt{3}}{3}$

8. $\frac{\sqrt{2}}{2}$

9. $\frac{\sqrt{2}}{2}$

10. $-\frac{\sqrt{3}}{2}$

11. -1

5.6 Half-Angle Identities

1. $-\sqrt{3}+2$

2. $-\frac{\sqrt{2-\sqrt{2}}}{2}$

3. $\frac{\sqrt{2-\sqrt{2}}}{2}$

4. $\frac{1}{2}$

5. $\frac{\sqrt{10}}{10}$

6. 9

7. $\frac{5\sqrt{26}}{26}$

8. $-\frac{\sqrt{50-15\sqrt{10}}}{10}$

9. $-\frac{3}{5}$

10. sin 15°

11. cos 40°

12. tan 25°

13. tan π

14. sin 100°

15. cos 275°

Find the exact value of each real number *y*.

1. $\sin^{-1}\dfrac{1}{\sqrt{2}}$

2. $\arcsin(1)$

3. $\tan^{-1}\left(-\sqrt{3}\right)$

4. $\cos^{-1}\dfrac{\sqrt{3}}{2}$

5. $\cos^{-1}(-\dfrac{\sqrt{2}}{2})$

6. $\arctan(-\dfrac{1}{\sqrt{3}})$

7. $\csc^{-1}(1)$

Give the degree measure of θ.

8. $\theta = \sin^{-1}\dfrac{1}{2}$

9. $\theta = \arccos(-\dfrac{\sqrt{3}}{2})$

10. $\theta = \tan^{-1}(1)$

Use a calculator to give each value in decimal degrees.

11. $\theta = \arcsin(0.1111111)$

12. $\theta = \cos^{-1}(-0.92170128)$

Give the exact value of each expression.

13. $\cos(\tan^{-1} 2)$

14. $\sec(\tan^{-1}\dfrac{12}{5})$

15. $\tan(\cos^{-1}\dfrac{3}{5})$

16. $\sec^{-1}(\tan\dfrac{3\pi}{4})$

Write each expression as an algebraic expression in *u*, *u* > 0.

17. $\csc(\sin^{-1}\dfrac{1}{u})$

18. $\cos(2\cos^{-1} u)$

Solve each equation for exact solutions in the interval $[0, 2\pi)$.

1. $2\cos x + 1 = 0$

2. $2\sin^2 x - 3\sin x + 1 = 0$

3. $\sqrt{2}\cos x + 1 = 0$

4. $4\csc^2 x + 3\csc x - 1 = 0$

5. $\sec x - \sqrt{2} = 0$

6. $2\sin^2 x = \sin x$

Solve each equation for exact solutions in the interval $[0°, 360°)$.

7. $(\sin\theta - 1)(2\sin\theta - 1) = 0$

8. $\sec\theta \cdot \csc\theta - 2\csc\theta = 0$

9. $\sin^2\theta = 1$

10. $\tan^2\theta = \tan\theta$

11. $2\sin^2\theta - \sin\theta - 1 = 0$

12. $4\sin^3\theta - 3\sin\theta = 0$

13. $\sqrt{3}\tan^2\theta - \tan\theta = 0$

14. $4\cos^2\theta + 4\sin\theta = 5$

15. $2\sin^2\theta - 3\sin\theta + 1 = 0$

16. $\sin\theta = \tan\theta$

Determine all solutions of each equation in radians (for x) to four decimal places.

17. $\cos^3 x = \cos x$

18. $2\sin^2 x \cdot \cos x - \cos x = 0$

19. $3\sec x + 7 = 0$

20. $2\cos x + \sin x + 1 = 0$

Solve each equation for exact solutions in the interval [0,2π).

1. $\sin 2x = \frac{\sqrt{3}}{2}$

2. $\cos^2 4x = 1$

3. $\sin 2x = \sqrt{3} \sin x$

4. $\tan 3x = 1$

5. $\sin 5x = 1$

6. $\cos 2x = -\frac{\sqrt{3}}{2}$

7. $\cos x = \cos \frac{x}{2}$

8. $\sin 2x + \sqrt{2} \cos x = 0$

9. $\cos 2x - \cos x - 2 = 0$

10. $\cos x \sin x = \frac{\sqrt{3}}{4}$

Solve each equation for exact solutions in the interval [0°, 360°).

11. $\cos^2 2\theta - 2\sin 2\theta + 2 = 0$

12. $\cos 2\theta \cdot \cos\theta - \sin 2\theta \cdot \sin\theta = \frac{1}{2}$

13. $\tan\frac{\theta}{2} = \sin\theta$

14. $\cos\theta = \sin\frac{\theta}{2}$

15. $\sin 3\theta = \sin\theta$

16. $16\sin^4 2\theta = 9$

17. $\sqrt{3}\sin\frac{\theta}{2} = -\frac{3}{2}$

18. $\sin\theta + \sin 3\theta + \sin 5\theta = 0$

19. $\sin 2\theta = \sqrt{3}\sin\theta$

20. $\cos 8\theta \cos 5\theta + \sin 8\theta \sin 5\theta = 0$

Solve each equation for *x*.

1. $y = \sec(2x) + \sqrt{5}$

2. $y = 10\sin x$

3. $4y = \tan(7x)$

4. $y = 5\cos(4x)$

5. $y = 5\sin\frac{x}{3}$

6. $y = -7\csc(3x)$

7. $y = \sin(x-1)$

8. $y = \tan x + 5$

9. $y = 3\cos x - 5$

10. $y = 3 - 7\tan x$

Solve each equation for exact solutions.

11. $6\sin^{-1}(\frac{x}{8}) = \pi$

12. $2\sin^{-1}(\frac{\pi+6}{x}) = \pi$

13. $\tan^{-1}(\frac{12}{5}) = \cos^{-1} x$

14. $\sin^{-1}(\frac{2}{\sqrt{5}}) = \tan^{-1} x$

15. $\cos^{-1} \frac{3}{\sqrt{10}} = \sin^{-1} x$

Solve each equation for exact solutions.

16. $\sin^{-1}\frac{1}{2} + \cos^{-1} x = \pi$

17. $\sin^{-1} x - \tan^{-1}\sqrt{3} = \frac{\pi}{6}$

18. $\cos^{-1} x + \tan^{-1} x = \frac{3\pi}{4}$

19. $\sin^{-1} x - \cos^{-1} 2x = \frac{\pi}{6}$

20. $\sin^{-1}(-\frac{1}{2}) + 2\tan^{-1} x = \frac{\pi}{2}$

6.1 Inverse Circular Functions

1. $\frac{\pi}{4}$ **2.** $\frac{\pi}{2}$ **3.** $-\frac{\pi}{3}$ **4.** $\frac{\pi}{6}$

5. $\frac{3\pi}{4}$ **6.** $-\frac{\pi}{6}$ **7.** $\frac{\pi}{2}$ **8.** 30°

9. 150° **10.** 45° **11.** 6.37937° **12.** 157.176°

13. $\frac{\sqrt{5}}{5}$ **14.** $\frac{13}{5}$ **15.** $\frac{4}{3}$ **16.** π

17. $\frac{2\sqrt{5}}{5}$ **18.** x **19.** $2x^2-1$

6.2 Trigonometric Equations I

1. $\frac{2\pi}{3}, \frac{4\pi}{3}$ **2.** $\frac{\pi}{6}, \frac{\pi}{2}, \frac{5\pi}{6}$ **3.** $\frac{3\pi}{4}, \frac{5\pi}{4}$

4. $\frac{3\pi}{2}$ **5.** $\frac{\pi}{4}, \frac{7\pi}{4}$ **6.** $0, \frac{\pi}{6}, \frac{5\pi}{6}, \pi$

7. 30°, 90°, 150° **8.** 60°, 300° **9.** 90°, 270°

10. 0°, 45°, 180°, 225° **11.** 90°, 210°, 330° **12.** 0°, 60°, 120°, 180°, 240°, 300°

13. 0°, 30°, 180°, 210° **14.** 30°, 150° **15.** 30°, 90°, 150°

16. 0, 180° **17.** $\frac{n\cdot\pi}{2}$ **18.** $\frac{\pi}{4}+\frac{n\cdot\pi}{2}, \frac{\pi}{2}+n\cdot\pi$

19. $2.0137+2n\pi, 4.2695+2n\pi$ **20.** $2.4981+2n\pi, 4.724+2n\pi$

6.3 Trigonometric Equations II

1. $\frac{\pi}{6}, \frac{\pi}{3}, \frac{7\pi}{6}, \frac{4\pi}{3}$

2. $0, \frac{\pi}{4}, \frac{\pi}{2}, \frac{3\pi}{4}, \pi, \frac{5\pi}{4}, \frac{3\pi}{2}, \frac{7\pi}{4}$

3. $0, \frac{\pi}{6}, \pi, \frac{11\pi}{6}$

4. $\frac{\pi}{12}, \frac{5\pi}{12}, \frac{3\pi}{4}, \frac{13\pi}{12}, \frac{17\pi}{12}, \frac{7\pi}{4}$

5. $\frac{\pi}{10}, \frac{\pi}{2}, \frac{9\pi}{10}, \frac{13\pi}{10}, \frac{17\pi}{10}$

6. $\frac{5\pi}{12}, \frac{7\pi}{12}, \frac{17\pi}{12}, \frac{19\pi}{12}$

7. $0, \frac{4\pi}{3}$

8. $\frac{\pi}{2}, \frac{5\pi}{4}, \frac{3\pi}{2}, \frac{7\pi}{4}$

9. π

10. $\frac{\pi}{6}, \frac{\pi}{3}, \frac{7\pi}{6}, \frac{4\pi}{3}$

11. 45°, 225°

12. 20°, 100°, 140°, 220°, 260°, 340°

13. 0°, 90°, 270°

14. 60°, 300°

15. 0°, 45°, 135°, 180°, 225°, 315°

16. 30°, 60°, 120°, 150°, 210°, 240°, 300°, 330°

17. none

18. 0°, 60°, 120°, 180°, 240°, 300°

19. 0°, 30°, 180°, 330°

20. 30°, 90°, 150°, 210°, 270°, 330°

6.4 Equations Involving Inverse Trigonometric Functions

1. $x = \dfrac{\sec^{-1}(y-\sqrt{5})}{2}$

2. $x = \sin^{-1}\dfrac{y}{10}$

3. $x = \dfrac{\tan^{-1}(4y)}{7}$

4. $x = \dfrac{\cos^{-1}(y/5)}{4}$

5. $x = 3\sin^{-1}\dfrac{y}{5}$

6. $x = \dfrac{\csc^{-1}(-y/7)}{3}$

7. $x = \sin^{-1} y + 1$

8. $x = \tan^{-1}(y-5)$

9. $x = \cos^{-1}\dfrac{y+5}{3}$

10. $x = \tan^{-1}\dfrac{3-y}{7}$

11. 4

12. $\pi + 6$

13. $\dfrac{5}{13}$

14. 2

15. $\dfrac{\sqrt{10}}{10}$

16. $-\dfrac{\sqrt{3}}{2}$

17. 1

18. -1

19. $\dfrac{1}{2}$

20. $\sqrt{3}$

Determine the remaining sides and angles of each triangle *ABC*.

1. $B = 107°$, $C = 11°$, $a = 0.9$
2. $A = 29.7°$, $C = 92°$, $b = 16.4$
3. $A = 80°$, $B = 30°$, $b = 14$
4. $A = 54°$, $C = 54°$, $b = 25$
5. $A = 33°$, $C = 82°$, $b = 18$
6. $A = 19°$, $C = 102°$, $c = 37$
7. $B = 25°$, $C = 125°$, $b = 5$
8. $A = 43°$, $B = 71°$, $a = 215$
9. $A = 21.3°$, $B = 19.2°$, $a = 143$
10. $B = 60°$, $C = 60°$, $a = 5$
11. $A = 15°$, $B = 80°$, $b = 6$
12. $A = 39.5°$, $B = 83.8°$, $c = 63$
13. $A = 158.6°$, $C = 12.6°$, $a = 25$

Find the area of each triangle *ABC*.

14. $A = 110°$, $b = 6$, $c = 7$
15. $A = 5.25°$, $b = 4.5$, $c = 22$
16. $A = 62°$, $b = 0.97$, $c = 0.19$
17. $B = 58.3°$, $a = 9.6$, $c = 19.3$
18. $C = 70°$, $a = 27.6$, $b = 14$
19. $A = 54°$, $b = 25$, $c = 21.3$
20. $B = 65°$, $a = 10.8$, $c = 19.7$

Determine the number of triangles *ABC* possible with the given parts.

1. $A = 40°$, $a = 3$, $b = 5$
2. $B = 30.2°$, $a = 18.5$, $b = 12.2$
3. $A = 80°$, $a = 15$, $b = 25$
4. $B = 56°$, $a = 8$, $b = 8$
5. $A = 21°$, $a = 12$, $b = 31$

Find the unknown angles in triangle *ABC* for each triangle that exists.

6. $A = 60°$, $a = 10$, $b = 12$
7. $A = 29°$, $a = 53$, $b = 41$
8. $B = 33°$, $a = 38$, $b = 29$
9. $C = 49.6°$, $b = 23$, $c = 18$
10. $B = 16°$, $a = 8$, $b = 4$
11. $B = 75.3°$, $a = 6.2$, $b = 9.9$
12. $C = 129°$, $a = 10$, $c = 8$

Solve each triangle *ABC* that exists.

13. $A = 58°$, $a = 42.4$, $b = 50$
14. $A = 54°$, $a = 24$, $c = 17$
15. $C = 12°$, $a = 5$, $c = 5.4$
16. $A = 110°$, $a = 125$, $b = 100$
17. $A = 37°$, $a = 28$, $c = 40$
18. $C = 129°$, $a = 10$, $c = 8$
19. $A = 55°$, $a = 19$, $c = 18$
20. $B = 28.6°$, $a = 41$, $b = 53$

Solve each triangle.

1. $B = 43°$, $a = 91$, $c = 88$
2. $A = 120°$, $b = 3$, $c = 10$
3. $a = 10$, $b = 12$, $c = 14$
4. $A = 56°$, $b = 9$, $c = 3$
5. $a = 48$, $b = 75$, $c = 63$
6. $B = 10.5°$, $a = 6.8$, $c = 2.4$
7. $a = 30$, $b = 40$, $c = 55$
8. $A = 116°$, $B = 34°$, $c = 8.5$
9. $A = 26°$, $B = 64°$, $b = 8$
10. $C = 121°$, $a = 3$, $b = 3$
11. $a = 14.3$, $b = 30$, $c = 21$
12. $C = 15°$, $a = 6.25$, $b = 2.15$
13. $a = 80$, $b = 92$, $c = 124$
14. $A = 22°$, $B = 170°$, $a = 9$
15. $a = 4.4$, $b = 3.8$, $c = 5.2$

Find the area of each triangle *ABC*.

16. $a = 10$, $b = 12$, $c = 14$
17. $a = 48$, $b = 75$, $c = 63$
18. $a = 30$, $b = 40$, $c = 55$
19. $a = 4.4$, $b = 3.8$, $c = 5.2$
20. $a = 14.3$, $b = 30$, $c = 21$

Given vectors a and b, find - 4a and a + 3b.

1. $\mathbf{a} = 4i - 3j,\ \mathbf{b} = i - 2j$

2. $\mathbf{a} = \langle -2,3 \rangle,\ \mathbf{b} = \langle -2,1 \rangle$

Find the magnitude and direction angle for u.

3. $\langle 6,-6 \rangle$

4. $\langle 2,-1 \rangle$

For each of the following, vector v has the given magnitude and direction. Find the magnitudes of the horizontal and vertical components of v, if α is the angle of inclination of v from the horizontal.

5. $\alpha = 127°,\ |\mathbf{v}| = 5$

6. $\alpha = 297°,\ |\mathbf{v}| = 45$

7. $\alpha = 27°,\ |\mathbf{v}| = 5$

8. $\alpha = 45°,\ |\mathbf{v}| = 4$

Two forces act at a point in the plane. The angle between the two forces is given. Find the magnitude of the resultant force.

9. forces of 2.4 and 1.3 newtons, forming an angle of 120°

10. forces of 3 and 8 pounds, forming an angle of 80°

Given $\mathbf{u} = \langle -2,4 \rangle$ and $\mathbf{v} = \langle -2,3 \rangle$, find the following.

11. 6**u** + 2**v**

12. 2**u** + 3**v**

Find the dot product for each pair of vectors.

13. $\langle -1,2 \rangle,\ \langle 0,5 \rangle$

14. $3i - 3j,\ 2i + j$

Find the angle between each pair of vectors.

15. $\langle \sqrt{3},1 \rangle,\ \langle 3,3 \rangle$

16. $7i + 8j,\ -2i + 10j$

Solve each problem.

1. An airplane is traveling with an airspeed of 420 mph and a heading of 72°. A wind of 32 mph is blowing at a heading of 130°. Find the ground speed and the course of the airplane.

2. A 240-pound box is on a 36° ramp. Find the magnitude of the component of the force that is parallel to the ramp.

3. Find the horizontal and vertical components of a 3-pound force that makes an angle of 60° with the x-axis.

4. A boat with a ground speed of 16 mph in still water is crossing a river perpendicular to a current of 6 mph. By what angle is the boat pushed off its course?

5. An airplane is traveling with an airspeed of 270 mph and a heading of 75°. A wind of 72 mph is blowing at a heading of 165°. Find the ground speed of the plane.

6. A 4200-pound car is on a 25° ramp. Find the force required to keep the car from rolling down the ramp.

7. A force of 750 pounds is needed to pull a sailboat and its trailer up a ramp inclined 20° from the horizontal. Find the weight of the sailboat and trailer.

8. Find the magnitude of the resultant force when forces of 170 pounds and 150 pounds act at an angle of 50° to each other.

9. Refer to exercise 8 above. Find the angle between the resultant force and the 170-pound force.

10. Find the amount of force required to pull a 250-pound crate up a ramp with an incline of 30°.

11. An airplane is traveling with an airspeed of 400 mph and a heading of 320°. A wind of 70 mph is blowing at a heading of 190°. Find the ground speed of the airplane.

12. A boat with a groundspeed of 13 mph in still water is crossing a river perpendicular to a current of 7 mph. By what angle is the boat pushed off its course?

13. Find the magnitude of the resultant force when forces of 5 pounds and 7 pounds act at an angle of 50° to each other.

14. A boat with a ground speed of 15 mph in still water is crossing a river perpendicular to a current of 3 mph. By what angle is the boat pushed off its course?

7.1 Oblique Triangles and the Law of Sines

1. $A = 62°, b = 0.97, c = 0.19$

2. $B = 58.3°, a = 9.55, c = 19.26$

3. $C = 70°, a = 27.57, c = 26.31$

4. $B = 72°, a = 21.3, c = 21.3$

5. $B = 65°, a = 10.8, c = 19.7$

6. $B = 59°, a = 12.3, b = 32.4$

7. $A = 30°, a = 5.9, c = 9.7$

8. $C = 66°, b = 298, c = 288$

9. $C = 139.5°, b = 129.5, c = 255.7$

10. $A = 60°, b = 5, c = 5$

11. $C = 85°, a = 1.6, c = 6.1$

12. $C = 56.6°, a = 48, b = 75$

13. $B = 8.75°, b = 10.4, c = 15$

14. 19.7

15. 4.5

16. 0.08

17. 78.8

18. 181.5

19. 215.4

20. 96.4

7.2 The Ambiguous Case of the Law of Sines

1. none

2. one

3. none

4. one

5. two

6. none

7. $B = 22.03°,\ C = 128.97°$

8. $A_1 = 45.5°,\ C_1 = 101.5°,\ A_2 = 134.5°,\ C_2 = 12.5°$

9. $A = 53.5°,\ B = 76.8°$

10. $A_1 = 33.5°,\ C_1 = 130.5°,\ A_2 = 146.5°,\ C_2 = 17.5°$

11. $A = 37.1°,\ C = 67.6°$

12. none

13. No Solution

14. $B = 91°, C = 35°, b = 29.66$

15. $A = 11.1°, B = 156.9°, b = 10.19$

16. $B = 48.74°, C = 21.26°, c = 48.23$

17. $B_1 = 22.3°, C_1 = 120.7°, b_1 = 17.6$ or $B_2 = 83.7°, C_2 = 59.3°, b_2 = 46.2$

18. No Solution

19. $B = 74.1°, C = 50.9°, b = 22.3$

20. $A = 21.7°, C = 129.7°, c = 85.2$

7.3 The Law of Cosines

1. $A = 70.9°, C = 66.1°, b = 65.7$

2. $B = 12.7°, C = 47.3°, a = 11.79$

3. $A = 44.4°, B = 57.1°, C = 78.5°$

4. $B = 105.2°, C = 18.8°, a = 7.7$

5. $A = 39.5°, B = 83.8°, C = 56.6°$

6. $A = 163.9°, C = 5.6°, b = 4.5$

7. $A = 32.2°, B = 45.2°, C = 102.6°$

8. $C = 30°, a = 15.3, b = 9.5$

9. $C = 90°, a = 3.9, c = 8.9$

10. $A = 29.5°, B = 29.5°, c = 5.2$

11. $A = 25.6°, B = 115.1°, C = 39.3°$

12. $A = 157.4°, B = 7.6°, c = 4.21$

13. $A = 40.1°, B = 47.9°, C = 92°$

14. No Solution

15. $A = 56°, B = 45.7°, C = 78.3°$

16. 58.79

17. 1503.30

18. 585.47

19. 8.19

20. 136

7.4 Vectors, Operations, and the Dot Product

1. $-16i+12j$; $7i-9j$ **2.** $\langle 8,-12\rangle$; $\langle -8,6\rangle$ **3.** $6\sqrt{2}$, 315° **4.** $\sqrt{5}$, 333.4°

5. $-3,4$ **6.** 20.4, 40.1 **7.** 4.46, 2.27 **8.** $2\sqrt{2}, 2\sqrt{2}$

9. 2.1 newtons **10.** 9 pounds **11.** $\langle -16,30\rangle$ **12.** $\langle -10,17\rangle$

13. 10 **14.** 3 **15.** 15° **16.** 52.5°

7.5 Applications of Vectors

1. 438 mph, 75.6° **2.** 141 pounds **3.** $\frac{3}{2}, \frac{3\sqrt{3}}{2}$ **4.** 21°

5. 279 mph **6.** 1775 pounds **7.** 2193 pounds **8.** 290 pounds

9. 23.3° **10.** 125 pounds **11.** 359 mph **12.** 28°

13. 11 pounds **14.** 11°

Identify each number as real, complex, pure imaginary or nonreal complex.

1. 3

2. $4i$

3. $2-7i$

4. $\sqrt{7}$

5. $\sqrt{-49}$

Write each number as the product of a real number and *i*.

6. $\sqrt{-16}$

7. $\sqrt{-7}$

8. $-\sqrt{-36}$

9. $-\sqrt{-120}$

Multiply or divide, as indicated. Simplify each answer.

10. $\sqrt{-9}\sqrt{16}$

11. $\sqrt{-2}\sqrt{-32}$

12. $\dfrac{\sqrt{-48}}{\sqrt{12}}$

13. $\dfrac{\sqrt{30}}{\sqrt{-90}}$

14. $\dfrac{\sqrt{-8}\sqrt{-4}}{\sqrt{2}}$

Find each sum or difference. Write the answer in standard form.

15. $(4+5i)+(2+3i)$

16. $(7-6i)+(2-3i)$

17. $(2+5i)-(3+4i)$

18. $(4-3i)-(3-4i)$

Find each product. Write the answer in standard form.

19. $(5-2i)(3+4i)$

20. $(2+5i)^2$

21. $i(4-3i)^2$

22. $(1+i)(1-i)(2+3i)$

Simplify each power of *i*.

23. i^{28}

24. i^{29}

25. i^{-5}

Find each quotient. Write the answer in standard form.

26. $\dfrac{2+5i}{4+i}$

27. $\dfrac{-3+2i}{-1-4i}$

28. $\dfrac{-7}{-i}$

Find the sum of each pair of complex numbers.

1. $5-3i,\ -2i$
2. $-2+7i,\ -8-3i$
3. $8,\ -3i$
4. $3-i,\ 2i$
5. $3,\ 2+3i$
6. $-4-2i,\ -2+3i$

Write each complex number in rectangular form.

7. $6(\cos 210° + i\sin 210°)$
8. $5(\cos 315° + i\sin 315°)$
9. $2cis\ 120°$
10. $4cis\ 210°$
11. $2cis\ 150°$
12. $4cis\ 300°$
13. $7cis\ 135°$

Write each complex number in trigonometric form $r(\cos\theta + i\sin\theta)$ with θ in the interval [0°, 360°).

14. $\sqrt{3}+i$
15. $6i$
16. 7
17. $-1+\sqrt{3}i$
18. $3-3i$
19. $-2i$
20. $1+3i$

Find each product and write it in rectangular form.

1. $[2(\cos 150° + i\sin 150°)][3(\cos 0° + i\sin 0°)]$
2. $(8cis240°)(2cis300°)$
3. $(4cis0°)(2cis60°)$
4. $(3cis150°)(4cis240°)$
5. $(2cis180°)(3cis90°)$
6. $(2cis270°)(4cis90°)$

Find each quotient and write it in rectangular form.

7. $\dfrac{2cis120°}{4cis210°}$
8. $\dfrac{cis300°}{cis180°}$
9. $\dfrac{12cis60°}{3cis120°}$
10. $\dfrac{9cis30°}{5cis60°}$
11. $\dfrac{5cis60°}{2cis300°}$
12. $\dfrac{cis120°}{cis30°}$
13. $\dfrac{3cis30°}{6cis120°}$
14. $\dfrac{10cis270°}{3cis0°}$
15. $\dfrac{\sqrt{2}cis45°}{10cis225°}$
16. $\dfrac{2cis90°}{6cis135°}$

Use a calculator to perform the indicated operations. Give answers in rectangular form.

17. $(3cis18.75°)(6cis230.5°)$
18. $\dfrac{3cis18.75°}{6cis230.5°}$
19. $(3cis20°)(4cis40°)$
20. $\dfrac{12cis52°}{3cis110°}$

Find each power. Write each answer in rectangular form.

1. $[2(\cos 15° + i\sin 15°)]^4$

2. $[2(\cos 30° + i\sin 30°)]^5$

3. $[5(\cos 20° + i\sin 20°)]^3$

4. $[3cis40°]^3$

5. $(-1+\sqrt{3}i)^{12}$

6. $2(\sqrt{3}+i)^7$

7. $[cis0°]^{20}$

Find all cube roots of each complex number. Leave answers in trigonometric form.

8. $8(\cos 135° + i\sin 135°)$

9. $-27i$

10. -1

11. 27

12. $-2+2i$

13. $-\dfrac{125}{2}(1+\sqrt{3}i)$

14. $-4\sqrt{2}(1-i)$

Find all complex number solutions of each equation. Leave answers in rectangular form.

15. $x^3 - i = 0$

16. $x^3 + 8 = 0$

17. $x^3 - 27 = 0$

18. $x^4 + 1 = 0$

19. $x^3 - 64 = 0$

20. $x^4 - 81 = 0$

21. $x^4 - 16 = 0$

Convert the following points from polar coordinates to rectangular coordinates.

1. $(2, \frac{4\pi}{3})$

2. $(-\sqrt{2}, 135°)$

3. $(6, 210°)$

4. $(2, \pi)$

Convert the following points from rectangular coordinates to polar coordinates where $0° \le \theta < 360°$.

5. $(4, -5)$

6. $(-1, \sqrt{3})$

7. $(-3, 6)$

8. $(-1, 1)$

For each rectangular equation, give its equivalent polar form.

9. $x^2 + y^2 - 2x = 0$

10. $x + y = 4$

11. $x^2 + y^2 = 9$

Identify the type of polar graph.

12. $r = 2\sin\theta$

13. $r = 4\sin 2\theta$

14. $r = 2\cos\theta$

15. $r = \theta$

16. $r = 3 + 2\cos\theta$

For each equation, find an equivalent equation in rectangular coordinates.

17. $r = \sec\theta$

18. $r^2 = 9\sin 2\theta$

19. $r = 2\cos\theta$

20. $\theta = \frac{\pi}{3}$

For each plane curve find a rectangular equation for the curve.

1. $x = \sin t,\ y = \cos t$ for t in $[0, 2\pi)$

2. $x = t^2 - 1,\ y = t + 3$ for t in $(-\infty, \infty)$

3. $x = \dfrac{1}{\sqrt{t+1}},\ y = \dfrac{t}{t+1}$ for t in $(-1, \infty)$

4. $x = 3t - 2,\ y = t + 1$ for t in $(-\infty, \infty)$

5. $x = 2 + 3\cos t,\ y = 3 + 2\sin t$ for t in $[0, 2\pi)$

6. $x = t^2,\ y = -2t^2 + 1$ for t in $(-\infty, \infty)$

7. $x = \dfrac{1}{4}t,\ y = t^2$ for t in $(-\infty, \infty)$

8. $x = 7\sin t,\ y = 7\cos t$ for t in $(-\infty, \infty)$

9. $x = 3 + \sin t,\ y = \cos t - 2$ for t in $(-\infty, \infty)$

10. $x = t^2,\ y = t^3$ for t in $(-\infty, \infty)$

11. $x = \dfrac{1}{2}\sin^2 t,\ y = \cos^2 \mathrm{t}$ for t in $(-\infty, \infty)$

12. $x = t - 1,\ y = \dfrac{t}{t-1}$ for t in $(-\infty, \infty)$

13. $x = \tan t,\ y = 2\tan t + 3$ for t in $(-\infty, \infty)$

14. $x = 3 + 2\sec t,\ y = 2 + 4\tan t$ for t in $(-\infty, \infty)$

15. $x = 2 - t^2,\ y = 3 + 2t^2$ for t in $(-\infty, \infty)$

16. $x = 3\sin t,\ y = 4\cos t$ for t in $[0, 2\pi)$

17. $x = t^3,\ y = 3\ln t$ for t in $(0, \infty)$

18. $x = \sqrt{t},\ y = \sqrt{1-t}$ for t in $(0, 1)$

19. $x = 3t,\ y = 6t + 4$ for t in $[0, 2]$

8.1 Complex Numbers

1. real, complex

2. pure imaginary, complex, nonreal complex

3. complex, nonreal complex

4. real, complex

5. pure imaginary, complex, nonreal complex

6. $4i$

7. $i\sqrt{7}$

8. $-6i$

9. $-2i\sqrt{30}$

10. $12i$

11. -8

12. $2i$

13. $-\frac{\sqrt{3}}{3}i$

14. -4

15. $6+8i$

16. $9-9i$

17. $-1+i$

18. $1+i$

19. $23+14i$

20. $-21+20i$

21. $24+7i$

22. $4+6i$

23. 1

24. i

25. $-i$

26. $\frac{13}{17}+\frac{18}{17}i$

27. $-\frac{5}{17}-\frac{14}{17}i$

28. $-7i$

8.2 Trigonometric (Polar) Form of Complex Numbers

1. $5-5i$

2. $-10+4i$

3. $8-3i$

4. $3+i$

5. $5+3i$

6. $-6+i$

7. $-3\sqrt{3}-3i$

8. $\frac{5\sqrt{2}}{2}-\frac{5\sqrt{2}}{2}i$

9. $-1+\sqrt{3}i$

10. $-2\sqrt{3}-2i$

11. $-\sqrt{3}+i$

12. $2-2\sqrt{3}i$

13. $-\frac{7\sqrt{2}}{2}+\frac{7\sqrt{2}}{2}i$

14. $2cis30°$

15. $6cis90°$

16. $7cis0°$

17. $2cis120°$

18. $3\sqrt{2}cis(315°)$

19. $2cis270°$

20. $\sqrt{10}cis71.6°$

8.3 The Product and Quotient Theorems

1. $-3\sqrt{3}+3i$ **2.** -16 **3.** $4+4\sqrt{3}i$ **4.** $6\sqrt{3}+6i$

5. $-6i$ **6.** 8 **7.** $-\frac{1}{2}i$ **8.** $-\frac{1}{2}+\frac{\sqrt{3}}{2}i$

9. $2-2\sqrt{3}i$ **10.** $\frac{9\sqrt{3}}{10}-\frac{9}{10}i$ **11.** $-\frac{5}{4}+\frac{5\sqrt{3}}{4}i$ **12.** i

13. $-\frac{1}{2}i$ **14.** $-\frac{10}{3}i$ **15.** $-\frac{\sqrt{2}}{10}$ **16.** $\frac{\sqrt{2}}{6}-\frac{\sqrt{2}}{6}i$

17. $-6.38-16.83i$ **18.** $-0.43+0.26i$ **19.** $6+10.39i$ **20.** $2.12-3.39i$

8.4 De Moivre's Theorem; Powers and Roots of Complex Numbers

1. $8+8\sqrt{3}i$ **2.** $-16\sqrt{3}+16i$

3. $\frac{125}{2}+\frac{125\sqrt{3}}{2}i$ **4.** $-\frac{27}{2}+\frac{27\sqrt{3}}{2}i$

5. 4096 **6.** $-128\sqrt{3}-128i$

7. 1 **8.** $2cis45°$, $2cis165°$, $2cis285°$

9. $3cis90°$, $3cis210°$, $3cis330°$ **10.** $cis\frac{\pi}{3}$, $cis\pi$, $cis\frac{5\pi}{3}$

11. $3cis0°$, $3cis120°$, $3cis240°$ **12.** $\sqrt{2}cis45°$, $\sqrt{2}cis165°$, $\sqrt{2}cis285°$

13. $5cis80°$, $5cis200°$, $5cis320°$ **14.** $2cis45°$, $2cis165°$, $2cis285°$

15. $\frac{\sqrt{3}}{2}+\frac{1}{2}i$, $-\frac{\sqrt{3}}{2}+\frac{1}{2}i$, $-i$ **16.** $1-\sqrt{3}i$, $1+\sqrt{3}i$, -2

17. $3,\ -\frac{3}{2}+\frac{3\sqrt{3}}{2}i,\ -\frac{3}{2}-\frac{3\sqrt{3}}{2}i$

18. $\frac{\sqrt{2}}{2}+\frac{\sqrt{2}}{2}i,\ \frac{\sqrt{2}}{2}-\frac{\sqrt{2}}{2}i,\ -\frac{\sqrt{2}}{2}+\frac{\sqrt{2}}{2}i,\ -\frac{\sqrt{2}}{2}-\frac{\sqrt{2}}{2}i,$

19. $-2+2\sqrt{3}i,\ -2-2\sqrt{3}i,\ 4$

20. $3,\ 3i,\ -3,\ -3i$

21. $2,\ -2,\ 2i,\ -2i$

8.5 Polar Equations and Graphs

1. $(-1,-\sqrt{3})$

2. $(1,-1)$

3. $(-3\sqrt{3},-3)$

4. $(-2,0)$

5. $(\sqrt{41},308.7°)$

6. $(2,120°)$

7. $(3\sqrt{5},116.6°)$

8. $(\sqrt{2},135°)$

9. $r=2\cos\theta$

10. $r=\frac{4}{\cos\theta+\sin\theta}$

11. $r=3$

12. circle

13. rose

14. circle

15. spiral

16. limacon

17. $x=1$

18. $(x^2+y^2)^2=18xy$

19. $x^2+y^2=2x$

20. $y=\sqrt{3}x$

8.6 Parametric Equations, Graphs, and Applications

1. $x^2 + y^2 = 1$

2. $x + 1 = (y - 3)^2$

3. $y = 1 - x^2,\ x \geq 0$

4. $3y - x = 5$

5. $\dfrac{(x-2)^2}{9} + \dfrac{(y-3)^2}{4} = 1$

6. $y = -2x + 1, x \geq 0$

7. $y = 16x^2$

8. $x^2 + y^2 = 49$

9. $(x-3)^2 + (y+2)^2 = 1$

10. $y = \pm x^{3/2}$

11. $y = -2x + 1,\ 0 \leq x \leq \dfrac{1}{2}$

12. $y = \dfrac{x+1}{x}$

13. $y = 2x + 3$

14. $\dfrac{(x-3)^2}{4} - \dfrac{(y-2)^2}{16} = 1$

15. $y = -2x + 7,\ x \leq 2$

16. $\dfrac{x^2}{9} + \dfrac{y^2}{16} = 1$

17. $y = \ln x$

18. $y = \sqrt{1 - x^2}, x \geq 0$

19. $y = 2x + 4,\ 0 \leq x \leq 6$